The Art of War for Computer Security

Tom Madsen

The Art of War for Computer Security

 Springer

Tom Madsen
Atea A/S
Ballerup, Denmark

ISBN 978-3-030-28571-5   ISBN 978-3-030-28569-2   (eBook)
https://doi.org/10.1007/978-3-030-28569-2

This Springer imprint is published by the registered company Springer Nature Switzerland AG.
The registered company address is: Gewerbestrasse 11, 6330 Cham, Switzerland

Preface

Welcome to this book, *The Art of War for Computer Security*. Books on the ancient Chinese classic text *The Art of War* as applied to areas like politics and business have been written before, so why now one on computer security? Is it even applicable to computer security? I, of course, think yes. Otherwise why would I write this book? *The Art of War* is aimed at conflict between nations, and as we have seen, it can be applied to conflict between companies and political parties as well. The key word here is "conflict." Computer security is meant to protect information or information systems against compromise, from either accidents or malicious attacks. Malicious attacks against information or systems mean that we now have a conflict between the attacker and the victim organization, and this conflict can be governed, managed, and mitigated by using some of the points made in *The Art of War*, some 2500 years ago!

The Art of War is about, well, war! So, is this a book about this fancy new concept called cyberwar? No! I have worked for many years within the United Nations, and I have seen the consequences of war up close. War is not a concept which should be applied to anything beyond its original meaning. War on this, war on that. It dilutes the original meaning, and consequences, of war. But I am getting sidetracked; back to this book. The original book has 13 chapters, with individual points made within each of the chapters. There are many quotes from *The Art of War* in the present book, and these quotes are usually from points in the corresponding chapters. Now, not all of the chapters, or all the points therein, are applicable to computer security, so I have picked out the chapters and points from the original version that can be applied to computer security. So, if you have picked out this book as a guide to the full version of *The Art of War*, then I am sorry to say that I will disappoint you with this book. You will have to buy one of the many very good books out there covering the full version, something I highly recommend. You might even consider getting one of the versions aimed at one of the many other areas where *The Art of War* has been applied. These books can, and do, offer other insights into the finer points made in the original version, or even made in the version I have written here.

Who might benefit from reading this book? If you are a computer security professional trying to secure your organization's information systems against attacks, you will benefit from the points made. If you are a manager responsible for securing information systems, you will benefit from this book. If you are an attacker, hopefully a white hat kind of attacker, you will benefit from this book, especially the chapters covering the offensive areas of war. I know that this last point can be controversial, but, as defenders, we absolutely must know how an attacker could go about compromising our information systems! If we do not know how an attack might take place, we cannot defend against that attack. So, yes, there are points made in this book which are solely of benefit to an attacker.

I mentioned earlier that the original book has 13 chapters. I have chosen to largely maintain the original chapter structure. When I say mostly, it is because some of the chapters in the original cannot be applied to computer security. These chapters I have removed from this book. The chapters are "Maneuver," "Marches," and "Terrain." I have added a chapter as well. The first chapter in the book will give you an overview of who Sun Tzu was, and some of his history, at least as much as is known, to give you some of the context in which *The Art of War* was written those 2500 years ago.

Overview of Chapters

I begin all the chapters, except for Chap. 1, with a general introduction to the content within. After that introduction, I move on to the individual points made in the original version of *The Art of War* and relate these to computer security. Some chapters will be longer than others, since some areas are more applicable to computer security than others. All the chapters, aside from Chap. 1, have the original titles from *The Art of War*. The chapters in this book are:

- *Chapter 1: Who Was Sun Tzu?* Who was Sun Tzu, and how did it come about that he wrote *The Art of War*? In this chapter I will give you a little of the history surrounding Sun Tzu, and explain why I think some of the quotes from the original work can be applied to cybersecurity.
- *Chapter 2: Estimates.* This is one of the longer chapters of the book. "Estimates" in the context of computer security covers how you go about preparing your infrastructure and staff for the inevitable attack or compromise, and the various ways a motivated attacker might try to deceive you during an attack.
- *Chapter 3: Waging War.* Why are we continuously being attacked? What are the attackers after, and how should we best respond to an attack? This is one of the chapters where many of the points made are of equal benefit to both attackers and defenders.
- *Chapter 4: Offensive Strategy.* This might sound like a chapter for attackers only, but as defenders we have to know how an attacker might go about achieving his/her goals. We also need to know what an attacker is after, before we can mount an effective defense.

- *Chapter 5: Dispositions.* Here we are looking into how you arrange yourself and your organization in preparation for an attack. This covers things like training of staff, both IT and general staff, to ensure that they are aware of their responsibilities.
- *Chapter 6: Energy.* How do you conduct an attack or a defense with a minimum of energy and resources spent? In this chapter we look at how you go about securing your infrastructure in the most efficient manner. Cost is always a parameter for investments; this chapter will give you pointers on how to maximize the resources you have available.
- *Chapter 7: Weaknesses and Strengths.* In this chapter we look at the strengths and weaknesses we have as defenders, but in equal measure we will be looking at the same for an attacker. As defenders we must know where we are weak, to focus our defenses and become strong. An attacker has weaknesses and strengths as well, and if we know these, we can determine the most effective defenses.
- *Chapter 8: The Nine Variables.* In the original work, the nine variables are used as pointers to variables which need consideration before embarking on a war. In this chapter I relate these variables to computer security by pointing out areas where we must consider the fact that the situations we deal with are, well, variable. Computer security is a never-ending road, where new defenses are devised and new attacks to circumvent them are developed.
- *Chapter 9: The Nine Varieties of Ground.* This will be the shortest chapter of the book. We will be looking at how our infrastructure is related to the concept of ground, and how this ground can help us to be better defenders, or how this same infrastructure can aid an attacker.
- *Chapter 10: Attack by Fire.* When we as defenders are in the midst of responding to an attack, it always feels like we are running from one fire to the next. A skilled attacker will try to attack us under the radar, but if the attacker is going at us with all the bells and whistles, then the attacker will make sure to hide the real attack behind a lot of "noise" that we must respond to, to prevent us from mitigating the real attack.
- *Chapter 11: Employment of Secret Agents.* In this chapter it will become a little "cloak and dagger"-like. You could be forgiven for thinking that secret agents cannot be applied to computer security, but these kinds of agents have been used for industrial espionage for decades, and they are used by organized hacker groups to get inside their target organizations. The colleagues we have in our organizations are all potential secret agents, knowingly or not.
- *Chapter 12: The Final Word.* In this last chapter, I will try to consolidate the points I have made in the previous chapters and explain why I think the points I have raised in the individual chapters are important.

Ballerup, Denmark   Tom Madsen
July 2019

Contents

Who Was Sun Tzu?

I will begin this chapter with a quick introduction to the history of Sun Tzu. Please bear in mind that I am not a Sun Tzu scholar, and hence this introduction will by necessity be a light one.

Sun Tzu

By tradition, Sun Tzu is estimated to have been alive between 544 and 496 BC during the age of the Warring States. The Warring States was a time in China's history when seven individual nations were competing for supremacy. These nations were named Zhao, Qi, Qin, Chu, Han, Wei, and Yan. All of these nations were located in fertile eastern China. Sun Tzu was hired by the king of a smaller kingdom, Wu, to be the leader of his army in a struggle between Wu and neighboring Chu.

It was in the struggle between these two nations that Sun Tzu developed his theories of warfare that became the book *The Art of War*. One of the more famous stories about Sun Tzu is about when the king of Wu asked Sun Tzu to train some harem girls in warfare as a proof of his skills as a commander. Sun Tzu approached the test seriously and trained the harem girls, but when he asked a couple of the harem girls to conduct an exercise the girls just giggled. Sun Tzu thought that the girls might not have understood the instructions he had given, so he gave the instructions again to make sure the girls understood. Still the girls just giggled. Sun Tzu now was of the opinion that the girls did not take the exercise seriously, and so he killed the two girls. Harsh, but from then on, the harem girls, understanding the consequences of not going about these exercises seriously, went about them in a much more serious manner.

Figure 1 gives an overview of the states which existed during the time of the Warring States. Wu is on the bottom right, along the coastline, with Chu to the left. There are many stories about Sun Tzu himself and his actions during the war

© Springer Nature Switzerland AG 2019
T. Madsen, *The Art of War for Computer Security*,
https://doi.org/10.1007/978-3-030-28569-2_1

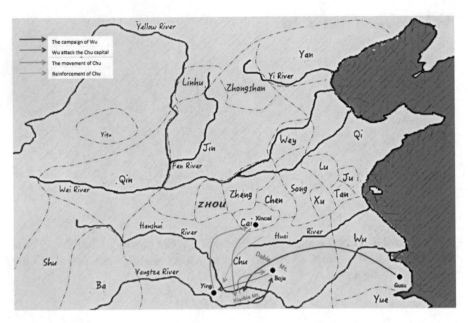

Fig. 1 Overview of the states which existed during the time of the Warring States (Source: Battle of Boju.png, https://en.wikipedia.org, by SY, used under CC BY 4.0)

between Wu and Chu. My personal favorite is one about when a commander in Chu's army chose to attack one of the states that were allied with Wu during the war. Instead of marching to the assistance of the ally, Sun Tzu chose to march on the capital of Chu. Why? Well, Sun Tzu knew through his spies that the commander of the forces allocated to the defense of the capital of Chu was in personal conflict with the commander of Chu's army in the field for power in the hierarchy of Chu. The commander of the army in the field did not want to give the defending commander in the capital the opportunity of gaining power by possibly defeating Sun Tzu's forces when they attacked the capital, so the field commander turned his troops around to march to the defense of Chu's capital. Thus, Sun Tzu aided his ally, but did not even engage his opponent in battle.

Eventually, Sun Tzu defeated Chu's forces and gained the king of Wu the territory of Chu as part of his kingdom. Sun Tzu then disappears from history aside from his book, *The Art of War* (Fig. 2). Now, in all fairness, there have been doubts expressed about whether Sun Tzu was an individual or a compilation of several individual people over the centuries. For our purposes in this book, however, we can ignore those doubts.

Fig. 2 *The Art of War* (Source: Bamboo book, binding, UCR.jpg, https://en.wikipedia.org, by vlasta2, used under CC BY 2.0)

Why Sun Tzu?

I touched a little on this in the introduction to this book; here, I would like to elaborate a little more. *The Art of War* has been applied to many different areas and purposes over the past century, with business being one of the more famous (infamous?) ones. The original work touches upon many different areas of warfare, warfare is a kind of conflict, and conflict exists in many contexts, not necessarily as physical in nature as war is.

In cybersecurity, we as defenders have an interest in keeping our infrastructures protected; the attackers on the other side have an interest in penetrating our infrastructures. That is a conflict of interest between us (the defenders) and the attackers. We can, in principle, be indifferent to the motivations of the attackers, be they political, money, or just being what is referred to as "trolls" in the lingo of hacking.

Trolls are being annoying just for the sake of being annoying, in case you are wondering.

Applying the various points raised in *The Art of War*, not all of which I can relate to cybersecurity, to conflict in various areas of life has been done by many different people in many ways over the years. I have chosen to relate some of these points to cybersecurity because many of the passages quoted from the original work are short and to the point and thus easily remembered. Admittedly, I have been cherry-picking quotes from the original work because I feel that relating some of the quotes to cybersecurity would have been more than a little contrived. In the Introduction, I mentioned that this book is not about cyberwar. I cannot emphasize this enough! If your main interest is cyberwar, there are plenty of books out there which focus on this topic exclusively.

My main focus in this book is the defense of our infrastructure. I will touch on areas and interpretations in *The Art of War* which can be applied if you are an attacker, but the main focus is on defense. Is the applicability of interpretations that are beneficial for attackers less important? No, but my role as a cybersecurity defender naturally predisposes me to look at defenses primarily. I am sure that you can come up with various scenarios and interpretations which can be used from an attacker's point of view. In fact, I really hope you do, but remember that if you do, then also think about how you would go about defending against such a scenario.

What's Next?

In the next ten chapters, I will be providing my comments and interpretations, as they relate to cybersecurity, for the quotes I have chosen from *The Art of War*. Each chapter will begin with a quick introduction to the subject, before getting to the quotes. Each chapter will also have a section with final remarks on what I hope you will take away from the chapter.

Does that mean that if you have taken a different lesson then you have misunderstood the content? No, it just means that your experience and viewpoint are different from mine; besides, having different takeaways for the quotes is a good thing. I do not claim any kind of authority over how to interpret *The Art of War* for cybersecurity defense. If your interpretation is different from mine, then good! Let's have a discussion about it; we might both learn something new.

I feel I should give you a quick introduction to the core of what I will be telling you about in the coming chapters. There will be three subjects I will be getting back to again and again and again ... Those three subjects are governance, risks, and threats. I am fully aware that there are plenty of other moving parts in any cybersecurity effort, but my personal feeling is that with those three subjects, any cybersecurity efforts will be based on a solid foundation which can be adapted to the ever-changing landscape of cybersecurity. But enough of this, let's get to the good stuff in the chapters with quotes from *The Art of War*.

Estimates

"Estimates" in the original work by Sun Tzu relates to the study of the intricacies of conflict and warfare. I will be equating estimates to a more holistic view of how to approach cybersecurity with a view to practical approaches and steps to be taken. The first quote in this chapter sums up nicely the importance of having a solid foundation in the basics of cybersecurity, as well as how these basics should be applied in the company or organization.

The "how" of going about studying cybersecurity in a company will depend on the kind of business or organization it is, as well as any regulatory or compliance requirements the company needs to meet. So, the sections in this chapter will by necessity be somewhat generic, but with enough detail for you to go about putting into place some of the recommendations.

Employees

Cybersecurity is not only a matter of having the right technology. I like to refer to cybersecurity as a 70/30 game, with the 30% being technology and the other 70% being people, processes, and procedures. Having the best technology in the world means nothing if the employees administering it do not have the right skills or the right policies and procedures in place surrounding the technology.

Many companies, especially those under some form of regulatory pressure, have policies and procedures in place, but these policies and procedures are hiding on a shelf somewhere and nobody knows where, except the people producing them for the external auditors. Policies and procedures have value beyond keeping the auditors happy, however! Business processes depend on procedures in a big way, but business processes are more than just the processes keeping the company running under normal circumstances. The processes activated during a cybersecurity crisis need procedures as well, and it is these procedures I am referring to here. The

© Springer Nature Switzerland AG 2019
T. Madsen, *The Art of War for Computer Security*,
https://doi.org/10.1007/978-3-030-28569-2_2

policies and procedures that keep a company or organization running on a day-to-day basis are well known to staff, even if the staff do not know where the policies and procedures are stored.

During a Crisis

It is during a real-life crisis that the policies, procedures, and business processes for dealing with a crisis become important to the company. And it is during such a crisis that any weaknesses in these policies and procedures reveal themselves. Do the staff who have to handle the crisis know where these policies and procedures are? Have the staff been trained in these procedures? Is there a dedicated organizational structure in place for dealing with the crisis? Have employees been given the authority to make decisions to deal with the crisis, or does everything have to be sent up the hierarchy to be dealt with, thus delaying the process? These questions must be dealt with before a cybersecurity crisis strikes a company. When the policies and procedures have been developed in such a way, the employees must then be trained in implementing these policies and procedures. Without regular training in these policies and procedures, the company will only have a paper tiger on a shelf somewhere with no real value during a real crisis. I consciously used the term "shelf" for storage of the policies and procedures for dealing with a cybersecurity crisis. There is no guarantee that you will be able to access your IT systems during a real crisis, so these policies and procedures must be available to staff offline.

Company Resiliency

It is not just the IT staff who need regular training in cybersecurity; the staff dealing with the day-to-day issues of the company need training in cybersecurity as well. Why? It is these staff who see the various kinds of phishing emails and scams coming into the email inboxes of staff worldwide on a regular basis. Keeping staffers aware of the risks of clicking on links in an email, for instance, is necessary to ensure that these staffers are continuously vigilant and skeptical of emails coming from unknown people or sources. Remember that these staffers are the first responders to suspicious emails, or behavior of computers, so making sure that they are well trained is key to having resilient cybersecurity in place. Training, as it relates to a real cybersecurity crisis, for these staffers must be in how to deal with the consequences of the crisis. If it is just the Internet connection that is down, then a solution might be to send the staff home to work from there, unless the systems they need for work are running on the premises and the Internet connection being down prevents them from working from home. Having procedures in place, as well as alternative business processes, is key to keeping the company running during a cyber crisis and, of course, having the staff trained in the alternatives will make sure that any disruption to the company or organization is kept to a minimum.

I am fully aware that I have only scratched the surface of all the issues surrounding the 70% dealing with the softer topics around the subject of cybersecurity. At the end of this chapter, I will suggest some frameworks and standards which can help you implement or structure your approach to cybersecurity.

Technology

Now I will turn to the 30% that technology contributes to cybersecurity. Technology is a tool in cybersecurity and can in no way be regarded as the solution. Technology can assist us in keeping our companies or organizations secure but cannot do so alone. For many years now, cybersecurity has been viewed as a question of having enough technology in place, whether the technology is hardware or software. This is a view which has been peddled by many of the security vendors in order to sell more products. Their doing so is understandable, but, as we have seen in the past decade, technology cannot stand alone but must be supported by softer checks and balances, as seen in many IT security frameworks and governance recommendations. Which technology platform a company or organization chooses for the technology stack is less important than how the company goes about implementing security in the technology stack.

Technology is an absolutely integral part of any cybersecurity implementation but, as mentioned previously, it only contributes some 30% to any cybersecurity implementation. How the technology is being managed and implemented will be a key contributor to the effectiveness of the technology as well. Have you outsourced the technology management to a third party vendor? If you have, then have you included this third party vendor in your cyber crisis management processes? This brings us back to some of the issues discussed in the "Employees" section of this chapter. Either the technology we choose must be known already by the employees, or training of the employees must be considered as a core part of the technology delivery. I will mention later, under one of the quotes, that many companies are reluctant to train their staff, for fear of losing them to another company or a security vendor, but having well-trained staff is an absolute requirement for a company or organization to have a security-resilient infrastructure and employees. Just imagine the chaos that could happen during a live incident if nobody knows who is responsible or what to do during the incident. Any third party vendors managing any part of the infrastructure must be included in these procedures; without that inclusion, a third party vendor might not know that there is a live crisis going on elsewhere in the infrastructure or what to do in such cases.

Quotes

War is a matter of vital importance to the state; the
   Province of life and death; the road to survival or ruin.
   It is mandatory that it should be thoroughly studied

If we substitute the word "state" with the word "company" and the word "war" with "computer security," then this quote becomes directly applicable to the modern company. Computer security is of vital importance to any company whose business relies on computers in any form. These days, any kind of company will be relying on computers in some shape or form; even production companies will be using computers to track production, pay bills, or manage the payment of salaries to employees.

In the "old" days, risk management meant evaluating risks involving investments and the like. Now risk management must include risks to IT systems. Assessing risks is becoming an increasingly complex endeavor, because IT systems, and the regulatory and compliance environment which companies are moving in, are increasing in complexity. Complexity also means that the configuration and development of IT systems aimed at functioning in a real environment become increasingly difficult. This difficulty increases the chances of misconfiguration or code errors, which attackers are utilizing with increasing frequency. Hardly a day goes by without some sort of computer breach being reported in the media.

I cannot overstate the importance of elevating the management of risks to IT systems to the board level. By risks I mean all risks to IT systems, not just the threat from hackers. IT is a core part of any business, and hence risk assessments must include integrity and availability as well. In computer security, we work with what is called the CIA triangle. The C stands for confidentiality, the I for integrity, and the A for availability. Any risk assessment of an IT system must include all three of these.

Risks and threats are continually evolving, hence the word "thoroughly" in the quote above. Aside from the regulatory compliance rules continually changing, security vendors are coming out with new ways of protecting our networks and IT systems all the time. The other side of the coin is that the attackers are continually working out ways of circumventing these same new ways of protecting ourselves. So, it boils down to an arms race between the defenders of our IT systems and the attackers. As defenders and companies, we must continually respond to new ways of attacking IT systems; make no mistake, the attackers are just as creative in developing their attacks as we are in finding new ways of defending ourselves! Computer security must be continually studied; by study here I do not mean formal university studies, although those kinds of study are important, but study of our infrastructure and the ability of that infrastructure to withstand new kinds of attack. Computer security is a continual journey toward an ever more elusive goal.

Which has the better trained officers and men;

A robust defensive posture of our IT systems relies not just on technology but, just as importantly, on the staff defending these IT systems. Computer security hardware and software are important, but without trained staff using these tools, the tools become meaningless. On top of this there are the policies and procedures in companies; are the staff made aware of and trained in these policies and procedures? Does the company even have such policies and procedures in place? By policies and procedures, I mean documents describing how the business will respond to an attack on its IT systems, or a business continuity plan. The business continuity plan might be activated by an attack on the IT systems but might equally well be invoked by a loss of power at the main company location.

Having staff trained in these policies and procedures is of vital importance, otherwise these policies and procedures might never be used when a situation arises mandating that they be activated. Having these policies and procedures in place is not just for getting a good review whenever there is an external audit. These policies and procedures are a core part of getting back to normal business operation whenever there has been an incident requiring the activation of them.

Having trained staff administering the security tools, whether they be hardware or software, is crucial to the overall security posture of a company. Just as important is having the same staff trained in the policies and procedures on computer security, whenever these policies and procedures are activated. Training in the policies and procedures does not require a big investment, nor an actual activation of a failover to an alternative datacenter. It can just be a desktop exercise, moving Post-its around a desk. The important part is that the staff involved in an incident requiring the activation of a policy or procedure know what they are responsible for and who to escalate any issues to.

I know there are companies saying that if they have staff who are well trained, they might lose those staffers to one of the security companies in the market. That is true, but what happens if we do not have well-trained staff? The well-trained staff should not just be the IT people who have to respond to attacks and the like—it is just as important that HR and finance people are trained to recognize a malicious email when they receive one. I will elaborate a little more on this under the next quote.

All warfare is based on deception

Whenever we receive a malicious email, the attacker is trying to trick us into doing something we should not. We have for decades now had to deal with phishing emails, and in more recent years the infamous CEO fraud attempts. The days are gone when fraudulent emails were badly spelled and easily recognized; the attackers are now spending a lot of time researching the companies they are attacking, in order to customize their emails to the audience in that specific company. But deception is not limited just to the various email attacks we are seeing on a daily basis.

Researching companies is now much easier than it was in the 1990s, for instance. Facebook and LinkedIn are excellent sources of detailed information on employees and company structures, allowing the targeting of malicious emails and their content to be extremely focused. The attackers know who reports to whom, as well as which functions the staff in the target company are performing. That is the reason for the successes in cases of CEO fraud; the deception is based on real knowledge of the target company and the employees working there.

As mentioned under the quote dealing with the training of staff, I would like to elaborate a little more on the end users here. It is not just the IT staff who need to be well trained; the end users are the ones seeing the attacks come in, in most cases. The end users must be continually trained to spot malicious emails and reminded not to click on links in these emails. Having the end users well trained and disciplined in email use will provide a company with a much more robust first line of defense. There are multiple technical solutions on the market, aimed at limiting the amount of spam and malicious emails in the inboxes of the end users, but none of these solutions can guarantee that none of these emails will reach the inboxes of the end users. Having well-trained users, along with technology, is the best way for companies to create a robust first line of defense.

Although emails remain a staple of the ever-evolving cyberattacks, there are many different forms such attacks can take. Distributed denial of service (DDOS) comes to mind as an example. DDOS attacks are most often used as a means of pressuring a company to pay a ransom for the attack to stop, but a DDOS attack can just as easily be used as a way of masking an intrusion while the staff at the victim company are dealing with the DDOS attack. Although most attacks come in via Internet connections, many companies have out-of-band management of their networks. Out-of-band management can be used by the company itself, or its service provider, to get its network up and running again, in cases where the administrative staff cannot access the equipment via the Internet. Most enterprise networking companies provide routers with the option of having a connection via a mobile or ADSL connection. These connections are attack vectors for the bad guys, just as much as the main Internet connection is. Often these connections are not secured as much as the main Internet connection is, and an attacker will not spend an inordinate amount of time attacking the main firewall if he/she can just walk in through the back door.

Attack where he is unprepared

As mentioned earlier in this chapter, the attackers will choose the path of least resistance when attacking a victim. Make no mistake, there are always weaknesses in our infrastructure and attackers willing to exploit these weaknesses. Try as we might, we cannot provide 100% security in our infrastructures, and the attackers are very skilled in finding the weaknesses. Our job as computer security professionals is to make their job as difficult as possible with the resources we have available. How do we go about finding out where we are unprepared? Many companies have asked security companies to conduct penetration testing of their infrastructure, but if we look at the 20 SANS Institute critical controls, penetration testing is the very last point on that list, as number 20!

Don't get me wrong—penetration testing is a good idea—it is just that there are many other controls on the SANS list which will provide an increased level of security, long before a penetration test should be on the agenda for a company. The first one on the list asks if you know what kind of hardware is running in your infrastructure. Do you? Do I? What about software—do you know what kind of software is running in your infrastructure? Hardware and software, knowing what kinds and what versions are running in the infrastructure, are the first two controls on the SANS critical controls list. Why? Because knowing this is the first step in securing the hardware and software components. If we have no idea something is running in the infrastructure, we cannot patch or otherwise maintain such components, thus making these components the area of our infrastructure where we are unprepared!

This brings me to the point of running software that is no longer supported by the various vendors. There are many and often valid reasons for continuing to run this kind of hardware/software, but we must be conscious of the risks in doing so. Do you have an old Windows 2003 server running in your datacenter? If yes, have you taken steps to limit the exposure of this server to the Internet, or maybe even segmented this server onto its own virtual local area network (VLAN) and controlled which kinds of traffic can reach this server? Back to one of the initial points—do you even know if you have a Windows 2003 server running in the datacenter?

Preparation requires insight; without insight we cannot secure our infrastructures, or allocate resources in a manner that provides security in any kind of optimized form. There are many ways of getting this insight; which one to choose depends on the business requirements for this insight.

Final Remarks

So, in this chapter I have been telling you about some of the various issues that have to be dealt with by any company or organization to implement a cyber-resilient infrastructure and have security-aware employees. Cybersecurity has long ago passed the hurdle of becoming a core part of business or organizational concerns. The amount of money lost every year to cybercrime alone has guaranteed that, but the increasing "cyberization" of society has made cybersecurity a front and center issue for citizens around the world. Companies which can show responsibility for the data they have on their customers, as well as strong IT governance structures, will be the ones that customers can trust and that investors can invest in and be reasonably sure that their investments are going to a company with strong cybersecurity and a good governance structure in place.

Next Steps

As promised earlier, I will now mention some additional resources that you can consult in order to dig deeper into the issues I have raised in this chapter. There are many and varied options for governance frameworks out there. Here I will promote only two of them, but it is well worth your time to look at more than the two I elaborate on here.

Governance

The two governance frameworks that I will promote here are ISO 2700x and COBIT 5. Why those two? Those are the ones I have experience with and am comfortable talking about—it does not mean that these two are the best out there. The best ones are the ones that are working for your company or organization! Hence my recommendation that you spend some time researching some of the other ones, such as CMMi or COSO.

Let's begin with ISO 2700x. The ISO 2700x series focuses exclusively on various security procedures and controls. Most of the cloud vendors have this ISO standard implemented in their infrastructures, with full certification from an external auditor. You do not have to get full certification to benefit from this standard, however; in addition, getting this certification and maintaining it is a costly affair. There are many moving parts to the ISO 2700x standard; most of them focus on ISO 27001, since this standard focuses on techniques and controls related to cybersecurity. It is worth looking at the rest of the standards in this series, because some of them focus on specific IT areas, such as IP telephony or virtual private networks (VPNs).

Next, COBIT. COBIT is developed and maintained by ISACA. The latest version is version 2019 and COBIT is a massive framework, with lots of different areas of applicability. You can use COBIT 2019 for information security or risk or governance, and lots of areas in between. The advantage of COBIT is that you can implement it in your organization by choosing only the parts of COBIT 2019 that are relevant to you and your situation.

Throwing yourself at implementing a COBIT 2019 framework, or the ISO 27001 standard for that matter, is not something to approach lightly. These frameworks/standards do have guidance on how to go about implementing them, but your organization will have to be fully committed to the implementation, because doing so will require resources and time.

Waging War

The waging of war has a somewhat negative connotation, but I would like to turn the original meaning of this to a more civilian purpose. So, for this chapter I will change the waging of war to how we go about conducting our cybersecurity defenses. I will be touching on the same topic in some of the other chapters, but I will elaborate on it in more detail in this one.

Cybersecurity defenses are a lot more than just the hardware and software we have installed to help us with our defenses. How we go about maintaining this equipment is just as important to the efficiency of the equipment. We have, for decades now, been told of the importance of keeping our software and hardware updated with the latest patches, yet still we see on a continuing basis that the breaches reported in the media happened because a server or a piece of software was missing a software patch, typically one that had been available for months before the breach. Microsoft has done a good job of automating the rolling out of new patches to Microsoft software, but it is still a challenge to keep third party software updated. Java and Flash have been infamous examples of this in the past few years. What about our routers, switches, and firewalls, are they not in scope for software or firmware updates? Absolutely! Unfortunately, these pieces of equipment are frequently forgotten in the patching process in companies and organizations.

Processes or Procedures?

In all fairness, it requires a stringent focus to keep track of the patches to hardware and software installed in a company. It also requires us to have knowledge of all the hardware and software installed in our infrastructure. Without that knowledge, we will be unable to know which software and hardware we should have a patching procedure for, and without such a procedure, the possibilities of unpatched hardware and software hiding somewhere in our infrastructure are too great. The Center for Internet Security has created what it calls the CIS Top 20 Controls. I have included a

© Springer Nature Switzerland AG 2019
T. Madsen, *The Art of War for Computer Security*,
https://doi.org/10.1007/978-3-030-28569-2_3

full list of these controls in the appendix, but the first two controls on this list are to have a full inventory of hardware in use and to have a full inventory of software in use. Why? Because, as I said earlier, without these inventories we cannot create processes and procedures to keep the inventories up to date with the latest patches.

Staff

The key, the absolute key, to any kind of cybersecurity defense is the staff involved in keeping the defenses running and responding to any and all incidents in the infrastructure. Making sure that these staff are well trained and motivated will contribute immeasurably to the cybersecurity posture of any company or organization. I will be detailing a little more of this in the first quote in the next section of this chapter.

"Resiliency" is a buzzword that has appeared in cybersecurity in the last few years, but the buzzword does have some truth to it. We might as well expect to become compromised at some point, so having resiliency in place to deal with the compromise will be key to recovering from the attack quickly, and the staff are a core component of resiliency. Technology will be part of resiliency too, but relying on the technology alone will handicap any efforts during an incident, because who will have to deal with the incident? The staff might depend on help or insight provided by the technology, but in the end the recovery effort will depend on the staff.

Quotes

> The reason troops slay the enemy is because they
>   Are enraged

This quote sounds a little dark, does it not? If we turn this quote around a little and apply it to the staff in companies responsible for defending networks and infrastructure, then the quote applies more to the motivation of the staff. The quote is still somewhat dark in nature when we look at the motivation of the attackers, but I will get to that part a little later. Having staff on your payroll who are motivated will provide you with a level of defense beyond what is possible with "just" staff and tools. Motivated staff will be looking at how to improve the state of affairs on a continuing basis and look into the root causes of incidents to find out why they happened. Do not discount the time taken to find out the root causes of an incident. I know that recovering from the incident is important, and should be prioritized as the first step, but finding out after the fact what happened will help in mitigating future incidents of that nature, thus increasing the overall security in the infrastructure.

Having motivated staff is not necessarily giving them a huge salary, although that might be part of it, depending on their skills and expertise. More importantly, it is making sure that their skills do not get out of date, by continually training them in new technology as it is developed. Any IT staff member provides value to the company because of his or her skills. IT is moving at such a fast pace that, without continually being trained, the IT staff will decrease in value for any company. The same goes for the cybersecurity staff employed at a company. Making sure that the staff are well trained is a core part of keeping them motivated and thus much more effective in their jobs.

Attackers can be enraged as well—just ask some of the companies which have been at the receiving end of hacker attacks from hacktivists. Hacktivists are less concerned with making money from their attacks, but the costs of attacks from hacktivists are just as devastating to a company as attacks from malicious black hat hackers. It is difficult to predict when some company action will trigger an Internet uproar and maybe trigger an attack from hacktivists, but part of any company's response plan for a hacker attack, regardless of the motivation behind the attack, should include multiple different responses depending on the kind of attack and how far into the infrastructure the attacker has gotten.

> They take booty from the enemy because they
>   Desire wealth

This one is simple. Although there still exist script kiddies in their mom's basement, the vast majority of attacks against companies are motivated by money. Just look at the popularity of the various cryptolocker viruses in the last few years. Even in cases where an attack has stolen research data, money is the motivation, because the attacker will save money by not having to do the research herself/himself or the company for whom they have stolen the data will not have to do the research. Hackers for hire can be found on the dark web, and you can be absolutely sure that companies are hiring these hackers to spy on their competitors. Industrial espionage has existed for centuries, and with the IT revolution you no longer have to physically break into a company to steal files. You can accomplish the exact same result from anywhere in the world, just with a computer. Again, money is the motivator here.

> Hence what is essential in war is victory, not prolonged
>   Operations. And therefore, the general who understands
>   War is the minister of the people's fate and arbiter of
>   The nation's destiny

This quote brings me back to one of my initial points. Hackers, just like businesses, are concerned with costs versus benefits. Hackers will choose the path of least resistance when attacking a company; why should an attacker spend weeks trying to circumvent a firewall, when all he/she has to do is write a convincing phishing email asking users for their username and password?

As defenders we have to use our resources in an intelligent manner and not, for instance, use $1 million to protect an asset worth only $100. So, cost versus benefit is a consideration we must apply to the risk analysis we are conducting on a continuing basis. We must create a defense that is "good enough"; what good enough is depends on the kind of company we are defending and the business the company is in. Some verticals have more rules and regulations than others, and these rules and regulations will have an impact on the risk analysis we are doing and on the kinds of defense mandated by the rules and regulations.

Final Remarks

So, this chapter has been about how we go about managing our technology and keeping the staff involved in the management of the technology well trained and motivated. Both of these points are key to providing any company or organization with an acceptable level of cybersecurity. What the acceptable level is will depend on the company, as well as on whatever compliance requirements might be in place in the business area the company is in. Just like a company, the hackers are in it for the money. If we are to protect those assets in our company which will provide an attacker with the most value, we must spend some time in identifying these assets and put a value on them. We can then use that value to allocate the right amount of resources and not spend more money protecting an asset than the asset is worth.

Offensive Strategy

"Offensive strategy" in this context does not mean us attacking the attackers. But having an offensive strategy will cover how we go about defending ourselves against those who attack us, by having our defenses in place to mitigate the kinds of attack we can foresee being launched against our infrastructure. In addition to this, counterattacking whenever we see an attack against our infrastructure will quickly become unsustainable, because the IP address we see an attack coming from will undoubtedly not be the IP address of the real attacker. There is a very real risk that our counterattacking will cause damage to a completely innocent person or company whose system has been compromised earlier and just used for an attack against us.

At the time I am writing this book, the German minister of defense has called for counterattacks in cases where an attack has been identified as being from a nation state. How to identify an attack as being from a nation state has not been detailed, and with the ever-present difficulty of attribution on the Internet I have difficulties in seeing how this can be done with any degree of precision. Add to this the risk mentioned above that the attack is coming from someone completely innocent because their systems have been compromised and are being used as jumping-off point for the attack on us. If counterattacks become an acceptable strategy for dealing with attacks on our infrastructure, my prediction is that this will quickly deteriorate to an all-out brawl in cyberspace, because everybody will feel justified in their counterattacks and innocents will be the ones caught in the middle. I cannot overstate the consequences if counterattacks become legal or accepted practice!

In this chapter I will be covering a lot of different factors contributing to the security of a company, so let's get started.

© Springer Nature Switzerland AG 2019
T. Madsen, *The Art of War for Computer Security*,
https://doi.org/10.1007/978-3-030-28569-2_4

Threat Assessment

Threat assessment and the risk assessment coming in the next section are key to any successful cybersecurity strategy. Of course, we want to develop an effective strategy and implementation of that strategy, but before we can do that, we have to identify the threats against our infrastructure. As mentioned in my comments on one of the quotes later in this chapter, we absolutely must know everything about our infrastructure before we can conduct an effective threat assessment. Along with our infrastructure, we also must consider what kind of business we are in and what kind of data we have on our systems. This all contributes to the threats we will identify.

Let's look at an example here, just to make this a little more concrete. We are a web shop and sell widgets to our customers around the world. Our customers can order widgets from us using our webpage and have the widgets sent to them all over the world. Our customers can pay by credit card at the time of ordering, or they can have 30-day credit and pay via bank transfer. This is a very high-level description of our business, but we can already identify some threats against our business. Let's look at the most obvious one first, the credit cards. If we are storing credit card details on our systems, then there is a threat against these credit cards from hackers. It is pretty much every week that we see in the news that some company has lost credit card details of its customers. So, there is a threat against our stored credit cards. What else? Well, unless our due diligence is in order, we run the risk of sending our widgets, with 30-day credit, to a company that does not exist, thus losing money in the process.

Conducting a threat assessment is a huge endeavor, make no mistake about this! The bigger the company, the bigger the effort required, and if the company has multiple business areas a threat assessment must be done for each business area. Many companies hire a consultant to conduct a threat assessment, and let this consultant do the entire assessment. Hiring in an experienced consultant is a good idea, but you absolutely must include staff from the business areas under assessment! The external consultant will be fully able to identify threats to the business, but threats can be against business processes as well, and your staff are familiar with these processes and can make sure that these are evaluated as well.

Risk Assessment

Why a section on risk assessment in a chapter about offensive strategy? Because if we are developing a strategy for our cyber defense, we have to develop it in relation to the risks we identify as relevant to our infrastructure and the threats we identified under the threat assessment. If we do not, at best the strategy will be ineffective, and at worst it might contribute to attackers getting access to the infrastructure. To conduct an effective risk assessment, we need to know what kind of hardware and software we have running in the infrastructure. If we do not have this insight, we will be unable to design a security solution which can be expected to be effective in protecting us against risks and threats, and we might inadvertently use more resources protecting an asset than the asset is worth to the organization.

Any serious risk assessment must include, besides the hardware and software, risks from the administrators and end users of the infrastructure. We all know about the kinds of damage that a normal user can unintentionally inflict on our systems, never mind what a malicious user might do, but an administrator can also do unintentional damage as well as intentional. The risks surrounding these scenarios must be considered as well. On top of this, there are the business needs which both systems and users are working to meet, as well as the processes aimed at fulfilling these needs. Risks to these business needs are covered, somewhat, under the above risks from administrators and end users, but risk is not necessarily about errors by users or a hacker attack. It is just as big a risk to the business that the infrastructure supporting it is no longer available or cannot be brought back up in a timely manner because of a corrupted backup.

Conducting a risk assessment requires both skill and experience, as well as deep insight into the business areas in which the risk assessment is done. Because of this, risk assessments are often done maybe only once a year, or maybe only when new systems or hardware are implemented in the infrastructure. This is bad practice, because the threat landscape is changing continually and something that was not even on the radar last month might suddenly become a major risk to the business. Remember that IT is continually changing, and so are the attacks. A full infrastructure risk assessment is a major project and not something that should be done on a quarterly basis, maybe not even on a yearly basis, but assessment of risks should be included in any IT project, and the risks identified in the project should be included in the document containing the risks identified against the overall infrastructure, including those from both end users and administrators.

Quotes

> For to win one hundred victories in one hundred
>> Battles is not the acme of skill. To subdue the enemy
>> without fighting is the acme of skill

If an attacker can gain access to our infrastructure without us becoming alert to the attack and stay in the infrastructure without us noticing the malicious presence, then the attacker has won a victory without us knowing we have lost. Make no mistake here, if we do not know that an attack has occurred successfully and a malicious user is present in our network, then we have lost. How do we go about decreasing the probability of an unnoticed successful attack? We cannot guarantee that no attack will be successful! 100% security does not exist in the real world; if an attacker is willing to allocate the time and resources, then the attacker will eventually be successful in gaining access to our infrastructure.

Recent studies by Mandiant, a cyber security company, have indicated that a successful attacker will be present in an infrastructure for about 146 days on average before being detected. Try imagining the amount of data an attacker could extract in that amount of time, or what our infrastructure could be used for during that time? Not a pleasant thought! If we cannot prevent a successful attack, then how do we go about making sure that we detect it as soon as possible? We have plenty of tools at our disposal here, but none of these will be enough on their own. For many years, intrusion detection system/intrusion prevention system (IDS/IPS) solutions have been sold as the solution, and these tools do contribute to a strong defense, but they are not enough on their own. In more recent years, security information and event management (SIEM) solutions arrived on the scene, as a way of centralizing logs from various systems, including IDS/IPS tools. This brings me to the point of logging in a company. Do you log? Are you using the logs? In most companies, the answer is yes to the first question and no to the second one.

It is the no answer to the second question that has made SIEM solutions so popular in recent years. With a SIEM solution, you can centralize your logging and automate a lot of the analysis of the logs. Implementing a SIEM solution is not an easy job, but after implementation you will have a tool which can monitor your network, servers, and clients for malicious activity and alert you to it. Maintaining a SIEM solution, and all of the surrounding technology, requires skilled staff, which again brings me back to a previous point. Good cybersecurity requires skilled staff! Skills will get out of date as time goes by, and hence it is of the utmost importance to continually train your cybersecurity staff; that way, it is also much easier to retain those staff, because they will feel appreciated and will be entertained while doing the job.

> Thus, what is of supreme importance in war is to
>   Attack the enemy's strategy

Every company has one or more strategies: one for getting more customers, another for increasing revenue, and so on. Unfortunately, very few companies have a strategy for cybersecurity, and without one there is a risk of investments in cybersecurity not providing the intended value of the investment. A cybersecurity strategy is not just having an IT security policy, although that is part of it—a cybersecurity strategy consists of all the things involved in providing a company with cybersecurity.

What things you could reasonably ask? Any cybersecurity strategy must include the technology in use, business continuity plans, incident response plans, and the staff involved (and not just the IT staff); the end users must be made an integral part of any security strategy. Governance must also be included—any self-respecting geek's eyes will glaze over at the mention of governance, but with the increased complexity of IT as well as the increasing level of rules and regulations surrounding any modern company, governance will only increase in importance in the coming years. Corporate governance has been important for decades, because of the various high-profile corporate fraud cases that have occurred. My personal prediction is that the companies with good governance in place, including IT governance, will be the companies which will garner the most respect, and thus also be the companies which will receive the most investment. Governance will become a much more important differentiator in the coming years than it is now.

Now, back to cybersecurity strategy. If you do not already have a cybersecurity strategy, it is time to create one. If you already have one, do you keep it updated as new developments or threats occur? If you have a cyber strategy created ten years ago, do you think it still applies to today's threat landscape? A cyber strategy must include both the soft and the hard areas of cybersecurity. By that I mean that the strategy should include both the various hardware and software solutions included in the infrastructure as security measures, and also governance and awareness programs. If your strategy includes only hardware and software solutions, then an attacker will go after your employees. If you already have a cybersecurity strategy, and it is updated, then having a look at it to do a risk-based analysis of how it goes about securing your organization is a good idea, and not just one time, but on a continuing basis.

A cybersecurity strategy must change when the infrastructure changes. If you change your firewall to another or a newer version, then the cybersecurity strategy must change to incorporate any and all new functionality, even if this technology is not utilized immediately. Why? Because even if you choose not to incorporate new functionality into your infrastructure, the strategy must include the why of it: if the reason you did not incorporate it changes later because of changes to the threat landscape, then you must update the strategy to reflect this and the strategy will maintain the history of changes.

Having a cybersecurity strategy, and keeping it updated and maintained, is of vital importance to a modern company that relies on IT in pretty much any way. Creating and maintaining a cybersecurity strategy is not just a job for a CISO (Chief Information Security Officer); it is a job for the CEO and board of the company. The CISO and IT department should provide input to the strategy, but the CEO and board are ultimately responsible for the creation and maintenance of a security strategy.

Next best is to disrupt his alliances

You could reasonably say that you are not aware of any alliances in your company, but if you switch the word "alliances" with "service providers," then this quote might make more sense. Most companies have outsourced one or more areas of their business to focus on the core areas of the company, and many of the outsourced areas are related to IT, thus making the companies these services are outsourced to an integral part of the overall cyber defense of the company.

If an attacker cannot penetrate your infrastructure, then he/she will try to gain access by compromising one of your service providers. Many companies are not considering the fact that outsourcing all or part of their infrastructure to a service provider requires them to adapt their business continuity and incident response plans to include their service provider. In addition to this, outsourcing all or part of the infrastructure introduces a new attack vector, namely the service provider. I am going to return to the subject of governance here. A company outsourcing its IT infrastructure will still be responsible for the security of that infrastructure—you cannot outsource responsibility. This fact necessitates that the company cybersecurity strategy, the incident response plan, the business continuity plan, and the general IT processes must include the service provider as a core part of these. Has your company even included the service provider in these processes?

These various plans and processes must include who is responsible for what and under what circumstances. These steps are especially important when a company is in the midst of responding to a live incident. If the plans and processes have not included this information, any response will be confused at best and ineffective at worst. What about the contract you have with the service provider? Does it even include any cybersecurity requirements? These requirements should include the requirements that the company has with respect to its infrastructure maintenance and development, but just as important are the requirements that the company has of its service provider.

The important part to remember here is that any outsourcing will require a company to update and adapt its processes to include the service provider in these processes. Without these steps, the company will be unable to respond to an incident in an effective manner. On top of this, requirements on the cybersecurity of the service provider should be included in the contract with the provider.

He whose ranks are united in purpose will be victorious

This is important! You cannot have responsibility for cybersecurity distributed among many different departments or people. Doing so will result in the same confusion and lack of efficiency as not including a service provider in the processes, as mentioned above. Governments are a prime example of this error, with multiple ministries and departments each responsible for their own areas of cybersecurity. I am sure you can think of examples from your own government.

Especially in a situation where your company is responding to a live event, having the responsibilities and processes handled by people designated for that situation will result in a less confused and more effective response. Having staff and an organization whose responsibility is cybersecurity will ensure that the area gets the required focus, as well as dedicated and trained staff. I cannot overemphasize this! And as an organization becomes bigger, this point becomes even more important.

There have been discussions for many years on where a cybersecurity department should be in an organizational diagram. My personal feeling is that it should report directly to the CEO and board. Why? Because it is par for the course that a cybersecurity department will be handling most, if not all, of the various IT-related compliance and regulatory requirements that any company has to deal with daily. Responsibility for these requirements lies with the CEO and board of the company, and since these requirements will only increase in the coming years, it makes sense to me to have the CISO and his/her people reporting to the CEO and board.

Therefore, I say: Know the enemy and know yourself;
    In a hundred battles you will never be in peril

This is one of the most popular quotes from *The Art of War*. Let's begin with knowing ourselves; here I am referring to knowing our IT setup and infrastructure. Having a complete and up-to-date picture of our infrastructure is of the utmost importance if we are to secure this infrastructure. If there is any shadow IT hidden somewhere without our knowledge, we cannot secure that area of the infrastructure. What about our patch management—do we know if the applicable patches have been applied across the infrastructure? That last point is dependent on us knowing how many applications we have running and what kind of applications they are. What about servers? How many do we have, and what are their operating systems? With virtualization software, it has become increasingly difficult to get an accurate number for this, because servers can be spun up and shut down depending on need and load without any operator intervention.

What about the adversaries? How do we get to know about them? Part of that answer comes from knowing about ourselves. What would a potential attacker be interested in, in our infrastructure? If we are a bank, that answer is easy. Money! But money is a core motivator for most attacks, so, looking at our own infrastructure, how would an attacker try to make money from that? Cryptolockers have been extremely popular in recent years, but there are other ways of making money from an attack. Does your company have any product secrets or research data you would like to keep safe? Hackers will happily steal that data and sell it to the highest bidder.

Threat intelligence promises to increase our security posture by providing insight into attacks happening elsewhere in the world before they reach us, allowing us to implement protections against these attacks. Threat intelligence, real threat intelligence that is, not the label put on almost anything security-related by the marketing departments of the various vendors, can provide a company with both increased security and insight into the company infrastructure. Any deeper treatment of threat intelligence is outside the scope of this book, but I recommend that you talk with your chosen vendor(s) about their offerings in this area. The takeaway for this quote is that you absolutely must have complete insight into your infrastructure and you must know why an attacker might compromise you and what his/her goals are for the attack.

> When you are ignorant of the enemy but know
> Yourself, your chances of winning or losing are equal

This quote is fairly obvious. If we have no idea of who might be interested in attacking us, or why and how, then our defenses will be much less effective. There will always be someone ready to attack us just for the kicks or for money, but here we are worried about more targeted attacks. Without insight into who might be interested in getting access to our infrastructure and their motivations for doing so, we cannot expect our defenses to hold.

This point is especially poignant for the various organizations providing services for intelligence services or the military. Who the opponent is is more easily identified in those cases!

> If ignorant both of your enemy and yourself,
> You are certain in every battle to be in peril

There are undoubtedly many organizations in this situation. There can be many reasons, both valid and invalid, for being in such a situation. One might be because of the merging of two IT organizations due to a merger; make no mistake here, hackers are looking for announcements of mergers, because they know that IT infrastructures in the midst of a merger will be in flux and thus provide many avenues of attack. If you find yourself in such a situation, making sure that the infrastructures are merged in a structured and speedy manner will be of the utmost importance.

If you are in this situation because the organization you work for is just taking chances and cannot be bothered about computer security, then I would recommend you find another job if you cannot convince the organization otherwise. Lastly, if you do not know your own infrastructure, then recovering from an attack will be difficult, since you do not know how the attacker came in or where he/she has been in the infrastructure.

Final Remarks

The whole concept of offensive strategy is easy to misunderstand, but I hope you have recognized the importance of having insight into ourselves and the infrastructure we are protecting. It is not just the hardware and software we must have insight into; the various business processes are just as important here. We will have hardware and software supporting these processes with either data or calculations, maybe both, and these business processes must be included in the insight we need to secure ourselves and our companies. For many years, cybersecurity has focused on the hardware and software running in our infrastructures, and these are important, but with the increase in regulations and compliance requirements the various business processes aimed at these regulations must be included in the cybersecurity strategy.

Cybersecurity strategy: I harped on about this under one of the quotes in this chapter, and I can hardly overstate the importance of having one. Developing a cybersecurity strategy will be a big job, and it needs to be based on the threat and risk assessments mentioned at the beginning of the chapter. Convincing a company to create such a strategy can be difficult, and arguing for it can be just as difficult. My recommendation is to argue for it using economic arguments; that way you will also be using a language that the business people will understand. How? Every company has one or more strategies; they have these to focus their investments, among other things, on the things which will assist them in reaching the objectives of these strategies. Developing a cybersecurity strategy will help the business to focus its cybersecurity investments where they will provide the most value to the business. On top of this, you can argue for a cybersecurity strategy as a means of being compliant, and retaining compliance, with all of the various rules and regulations in today's business environments.

IT governance, along with the various aspects of business governance hopefully in place, will assist a company or organization in satisfying its cybersecurity needs as well. Having good governance in place, besides showing investors and the public that a company is responsible, will provide the business with assurance that its processes are contributing to its overall security and that the processes include steps to take risks into account when making decisions. Good governance is especially important to organizations with a project-based structure, or which are managing projects on behalf of other organizations. For additional insight into governance for projects, I recommend that you look at the Project Management Professional (PMP) certification from the Project Management Institute or the Certified in the Governance of Enterprise IT (CGEIT) certification from ISACA.

Dispositions

Dispositions can mean many things to many different people; in this context, dispositions are about how you go about the job of securing your company or organization. How do you view cybersecurity? How do you go about doing risk assessments? How do you go about doing threat assessments? The how of all the various areas that are cybersecurity will have an impact on the efficiency and effectiveness of the various controls and technology we go about implementing to protect our companies and organizations against cyberattacks.

The quotes in this chapter will be fairly philosophical in nature and open to interpretations in many ways. I have no doubt that you will have interpretations of your own, based on your own experiences. That is a good thing! I make no claim to possess the final truth here, quite the opposite actually. So please go ahead and take away whatever interpretations work for you from this chapter!

Quotes

> Invincibility depends on one's self; the enemy's
>   Vulnerability on him

Invincibility in computer security is a tall order. In fact, no matter how hard we try, we can never achieve 100% security. What we can achieve is a level of security which will make an attacker ignore us and go and look for an easier target. We ourselves, and the companies we are working for, will be the source of this "invincibility." How we approach computer security and how well we know the infrastructure we are trying to defend will, along with the governance structures in place, decide how well our defenses will work against a real-world attack. So, that all depends on us and the procedures we have in place, but it is equally important to have these procedures and security measures in place where they will be most

© Springer Nature Switzerland AG 2019
T. Madsen, *The Art of War for Computer Security*,
https://doi.org/10.1007/978-3-030-28569-2_5

effective and protect the right assets. As mentioned under the quotes in the previous chapter, part of our invincibility depends on us "knowing ourselves."

Our enemy, the hacker, has vulnerabilities as well. Just like us, a hacker has preferences and will specialize in certain technologies or software. On top of this, the hacker will prefer certain kinds of technique when compromising a victim. That is also the reason why security companies sometimes can identify specific hacker groups or individuals as being behind attacks, by their "signature," the preferred way that these people do their dirty deeds. That is why our security tools can help us identify an attack as it happens, because the attack may use a known technique. Since attacks are evolving all the time, we absolutely must keep these signatures up to date.

> Therefore, it is said that one may know how to win,
>   But cannot necessarily do so

What does this mean, you ask? Is it just "dumb" luck if my defenses are working against an attack? No, it is not "dumb" luck, because we ourselves can create our luck! If we have done our risk assessments and threat assessments, then we can implement and manage our defenses in the most effective way for us.

Around this point is, of course, the whole subject of governance of our IT, secure configuration of our IT, keeping our IT systems patched, and so on. . . . So, yes, we are the creators of our own luck. We already know how to win, but for some reason there is always one (or more) of the steps we know we must take that never gets taken.

> Invincibility lies in the defense; the possibility of
>   Victory in the attack

This quote mirrors the one which started this section. It is our defenses, not our attacks, which will ensure our security. Our defenses will not be relying just on technology, but also in equal measure on the softer policies, procedures, trained staff, and governance structures we put in place. Too often companies will trust their cyber safety to technology and completely disregard all of the softer checks and balances around the technology.

> Anciently those called skilled in war conquered an
>   Enemy easily conquered

Being easily conquered does not sound like a good thing, does it? Unfortunately, unless we are prepared for an attack, we will be easily conquered. Not all the stories we read in the newspapers on a weekly basis are because a company has been attacked by a "sophisticated" hacker, no matter what the company is stating in its press releases. The clear majority of these companies were easily conquered. To become eligible to get into a newspaper because of a hacker attack, a company has to be of significant size—so, a company of a size where you would expect that the resources for cybersecurity would be present. To be fair, however, some of the companies that have been in the news because of a cyber incident are allocating significant resources to cybersecurity and were just the victim of the ever-present risk of human error.

The point I would like to make here is that cybersecurity requires resources to be effective, but resources alone are not enough, because we as humans are prone to mistakes. The technology we use in cybersecurity might be misconfigured, or we might forget to apply the latest patches to the equipment. Automation has been presented as the solution to human error for many years, but automation of such complex technology as cybersecurity software and hardware is incredibly difficult. Governance and structure surrounding the technology will assist in reducing human error to a level where any errors will no longer be catastrophic if they are exploited by a malicious attacker.

Final Remarks

So, this chapter has been a "fluffy" one, with lots of soft recommendations and interpretations surrounding cybersecurity. Please do not disregard these softer issues. We are all human and are therefore susceptible to making errors. Having checks, balances, and controls in place in our infrastructure will make our cybersecurity measures much more effective as well as make any investigations we must make after the fact much easier. Separation of duties has been exercised in banks for centuries; implementing similar separation between administrators for critical systems is a good idea. It will produce a more cumbersome process, I know, but it will also make it difficult for any single person to use his/her privileged access to defraud the company or organization. Have you and your company been doing any assessments of the criticality of individual systems, to assess whether or not such a process might make sense?

Energy

Energy—we all know the feeling of having none of it! "Energy" in the original book referred to how to go about managing the resources, including people, allocated to the conduct of war. In this chapter I will relate energy to how we utilize the resources available to us in an efficient manner.

Just as we know the feeling of having no energy, we know about lacking resources to do our job in cybersecurity. There are many reasons for lacking resources, some good, some not so good. Money is undoubtedly the primary concern when we are talking about resources, but we should keep in mind that time is a resource as well. The time we spend doing cybersecurity is a finite resource—just talk to anybody who has been involved in a cybersecurity incident. During a live incident, we are running ourselves ragged, trying to keep up with new information as it flows to us while we are investigating the incident. So, time and money. How do we manage these resources efficiently?

Resource Management

However much we would like to have infinite resources, time and money are going to stay finite for the foreseeable future. So, to do our job efficiently, we must be able to manage our time and the money we have been allocated for cybersecurity wisely. "Wisely" can mean many things in different situations; the important thing to keep in mind is the business need. The business need is the yardstick for all decisions we make regarding cybersecurity.

It is the business which is allocating resources for cybersecurity, hopefully with a view to the threats and risks facing the business, and the business expects us to use those resources to protect it against cybersecurity events in a manner which reflects the risks and threats facing the business. As cybersecurity defenders, we would always like to have more resources at our disposal and we might even need these additional resources, but before we can get these we need to convince the business of

© Springer Nature Switzerland AG 2019
T. Madsen, *The Art of War for Computer Security*,
https://doi.org/10.1007/978-3-030-28569-2_6

this need. Here we are getting into the somewhat hidden part of resource management.

Every business and organization will need some level of cybersecurity. This level will depend on the threats and risks facing the business or organization. As cybersecurity defenders, it is our job to make sure that the business or organization is fully aware of the threats and risks it faces in regard to cybersecurity and to ask for resources at a level at which they can be used to defend against these risks and threats. What I am trying to say here is that, as cybersecurity defenders, we need to be good communicators and negotiators. We have to negotiate for budget allocations but, just as important, we must be able to communicate the need for additional resources or money. Without this skill, any cyber defender will be perpetually frustrated or stressed trying to keep the company or organization defended with limited resources at his/her disposal.

Quotes

> Generally, the management of many is the same as
>   Management of few. It is a matter of organization

"Organization" is the key word here. Part of any effective organization is dedicated leadership and dedicated responsibilities. I have mentioned it elsewhere that, to be effective, the responsibilities for cybersecurity must not be shared between different departments or organizations. Doing so will only result in confusion and a distinct lack of efficiency. Organizational effectiveness depends on these responsibilities being defined and implemented, and the same goes for the responsibility for cybersecurity. Efficiency here requires that the organization recognizes the need for a dedicated unit within it to handle the responsibility for cybersecurity. Depending on the size of the organization, it might make sense to put the responsibility for cybersecurity within IT. In such a case, there should be dedicated personnel in IT whose responsibility is cybersecurity.

I fully understand that the chosen model will depend on the size of the organization and that a smaller organization will not have the resources to dedicate specific employees to this area, nor may it have an IT organization at all. In cases like that, depending on how the organization approaches its IT needs, it might make sense to outsource its cybersecurity needs, or have a partner with which it can spar over its needs. The important thing here is to remember that even if you outsource your cybersecurity needs, you will still be the one responsible for the cybersecurity of your company. You can only outsource the technical administration of cybersecurity, not the responsibility.

> The primary colors are only five in number but their
>   Combinations are so infinite that one cannot
>   Visualize them all

This quote from *The Art of War* is my favorite one! It can be applied to so many things in life that I am tempted to call it a catch-all quote. As applied to cybersecurity, however, the quote neatly boils down the challenge of cybersecurity to the fact that the people who attack us are creative in the methods and attacks they apply in their "craft."

The attacks are usually based on some standard attack profiles; it is the combination of these profiles that makes being on the defending side of cybersecurity such a challenge. On top of this, there is the new technology which gets added to our networks at an increasing rate; just think of the Internet of Things (IoT) or Supervisory Control and Data Acquisition (SCADA) systems (SCADA systems are the kinds of system used in controlling our water supply and electricity systems and the like). Attackers know just as well as us that new technology has not been in the "wild" long enough to become hardened against the kinds of attack we see in the real world. So, we must be cognizant of all the "old school" attacks as well as be aware of all the new possibilities presented by all the new technology popping up all the time. The challenge is compounded by us not knowing what kinds of attack are possible against the new technology.

Most of the new technology is tested against the known attack scenarios; it is when the new technology gets put together with other new technology as well as the old technology that the weaknesses show themselves. What is the solution here? I am not sure there is one, beyond being aware of the risks when putting new technology into our production networks and infrastructure.

> His potential is that of a fully drawn crossbow;
> His timing, the release of the trigger

Now I am going to be a little "cloak and dagger"-ish. We have, for the last decade or so, been told that various industrial production systems are vulnerable to attack, and they are; we saw the very real physical consequences an attack can cause with the Stuxnet attack on the Iranian nuclear facilities in Natanz, where many of the centrifuges used in the process for enriching uranium were destroyed by the Stuxnet infection. That was the first time that we know of where a state has consciously attacked another with a cyberweapon. Since then, we have heard many accusations going back and forth between various states, accusing each other of infecting systems critical for society to function, in preparation for some kind of conflict between the states. There is absolutely no doubt in my mind that states are in fact doing this.

How does this relate to the quote that began this comment? We, as defenders, are protecting the systems which get hijacked in these political maneuverings. We are trying to defend the systems against someone whose only goal in infecting us is to use our systems in a political game, or just in case a conflict between states gets out of hand. Our systems are now the fully drawn crossbow. I do not know how you feel about that; my take on this is that to hijack essentially civilian systems for this purpose is so far outside of what I can accept and what I would expect most civilian governments to accept, that it is baffling to me that no international agreement has been reached on conducting what is essentially war in cyberspace.

Final Remarks

I have tried to equate the concept of energy to resources in this chapter. What I have not touched upon is the fact that you can have too many resources at your disposal. What! Really? Yes, you really can. This kind of situation usually arises after a cybersecurity incident has made a business or organization aware of the real-world consequences that a security incident can have on a business and its customers. Our job as professional cybersecurity defenders is to not get carried away when resources start flowing our way. There will come a time, on the other side of the incident, when all the resources we needed during the incident will become a burden, either because we no longer need the additional staff, or because we now have hardware and software which is no longer needed.

So, if you are to keep the respect and trust of the business, make sure to keep the urge to acquire additional resources to a reasonable level.

Weaknesses and Strengths

This chapter will be one of the longer ones. Weaknesses and strengths, we all have them, and they are what will make or break our cybersecurity efforts. Knowledge of our weaknesses can be made into a strength if we can look at ourselves and our infrastructure in an objective manner. When we have this knowledge, we can allocate our resources in a more effective manner as well. There is no reason to focus on areas where we are already strong; the efforts should be on developing our weaknesses into strengths, and the weak areas of our infrastructure must be made strong.

Weaknesses

We will have two kinds of weakness in our infrastructures: the known ones and the unknown ones. The known ones we can, and should, do something about. The unknown ones, by their very nature, are more difficult to deal with, and it is usually these weaknesses that are the cause of cybersecurity incidents.

The weaknesses we know about we can turn into strengths, making some new area of the infrastructure into a weak area. Dealing with weaknesses is a never-ending journey—or is it? Remember my remarks on resources in the previous chapter. Resources must be allocated based on the risk and threat profile of the company or organization. At some point, the defenses we are putting in place will be good enough. I know that good enough is never good enough to a cybersecurity defender, but we must keep in mind that the company is in business to make money and overspending on cybersecurity will cut into its profits.

What is good enough changes from company to company and from risk profile to risk profile, and part of our job as cybersecurity defenders is to make sure that the business is aware of the threats and risks facing it and to use the resources we are given in a responsible manner.

© Springer Nature Switzerland AG 2019
T. Madsen, *The Art of War for Computer Security*,
https://doi.org/10.1007/978-3-030-28569-2_7

Strengths

Knowledge of our strengths, and the strengths in our infrastructures, is part of what provides us with insight into our weaknesses. Knowing our strengths gives us the freedom to allocate resources to shore up our defenses where they are weak.

There will always be the temptation to rest on our laurels in areas where we "know" we are strong, but if we give into this temptation, we run the very real risk of becoming blind to weaknesses when things change in our infrastructure. With the speed of change and the new trends towards agility, changes in our infrastructures will become more frequent and will without a doubt become a contributing factor to new weaknesses in those infrastructures. So, strengths can become weaknesses and weaknesses should become strengths.

Quotes

> Appear at places to which he must hasten; move swiftly
>   Where he does not expect you

Every hacker attack or malicious software infection is unexpected, but attacks and malicious software are designed to stay well under the radar, so when we do detect that we are under active attack, often an attack appears from an angle we did not expect. We are all of us concerned with protecting our infrastructure and deeply worried about any areas of the infrastructure we might have missed, or not considered as an attack vector. Those are the areas that a skilled attacker will go for.

What is the takeaway from this quote? I could try the standard answer: expect the unexpected! But that seems a little too much like a cop-out. Most of us have been in a situation where an attacker entered the infrastructure via that one thing we did not think about. To limit the risk of missing that one thing, we must use a process whenever we go about doing assessments on our infrastructure. If we go about doing things in an ad hoc way, there is a risk of us missing one or more areas during our assessment where the risk is high enough for us to need to try to mitigate it. Remember, a risk assessment contains risks itself, namely of our not doing it thoroughly enough or going about it in an unserious way.

> To be certain to take what you attack is to attack a
>   Place the enemy does not protect. To be certain to hold
>   What you defend is to defend a place the enemy does
>   Not attack

This quote ties in nicely with the previous one. There are many cases where an attack goes for "that one thing" mentioned above which we did not think about protecting or did not know was an actual weakness in our defenses. Hackers are very skilled in finding these weaknesses and exploiting them. We cannot be equally strong in all places—I will return to this point in a quote later—so how do we handle this? Technology can be part of the solution. Network monitoring and log management will be able to alert us if something new and unexpected is running in our infrastructure, and with those tools we can find out where it comes from and put in mitigating controls to shore up our defenses in that area.

This quote is also the place where I have chosen to address the proverbial elephant in the room: the users. No amount of technology we put in can protect us from the mistakes of the users. Remember, we are users as well! We might be less susceptible to clicking on links in emails, but it only requires us to be unguarded for a single moment before we make a mistake as well. Monday morning before the first cup of coffee comes to mind. We can "harden" the users by continually conducting awareness programs to remind them of the dangers of clicking on links in emails from strangers, but in any kind of email-based attack it only requires one user to click on the link for the attack to be successful.

Few people think about this, but the users are part of our defenses too. I know that it does not feel like it at times, but it is the users who see the various emails with the nasty links in them, and hence they are our first line of defense. If we give in to the temptation to punish the users for their mistakes, then they will become afraid to go to IT if they feel they have made a mistake, and we will lose our first line of defense. When a user approaches us with a concern, we must thank them and write to the rest of the users to tell them about the email the first user got and ask them to go to IT with any concerns they might have about an email they have seen or clicked on.

> The enemy must not know where I intend to give
>   Battle. For if he does not know where I intend to give
>   Battle he must prepare in a great many places. And
>   when he prepares in great many places, those I have
>   to fight in any one place will be few

This quote, and the next one, is where I will address the point raised in the previous quote, on being strong everywhere. We cannot, however hard we try, be strong everywhere. To be strong somewhere, we must ensure that the places we protect are the ones most critical to the business or organization. To determine those areas, we must involve the business and the business processes in the evaluation of critical business areas. Remember, IT systems can be used to support a business process and not just contain data which needs protection. A database containing credit card numbers will of course need to be locked down thoroughly, but what about the business process handling these credit card numbers? Or the systems, not the database, used in the business process handling these credit card numbers?

Criticality can be challenging to determine for any kind of system or data, but without having processes for this determination we will be unable to focus our defenses on the areas and processes where they will be most effective. Without this focus, we might try to be strong everywhere, which neatly brings me to my next quote.

> For if he prepares to the front his rear will be weak,
>   and if to the rear his front will be fragile. If he
>   prepares to the left, his right will be vulnerable and if to
>   right, there will be few to his left. And if he prepares everywhere
>   he will be weak everywhere

None of us cybersecurity defenders wants to be weak everywhere; we want to be strong everywhere. I have already mentioned the fallacy of that statement after the previous quote. As defenders, we must create a strong foundation of security, and build on that foundation with stronger security measures where these stronger measures are needed and make sense for the business or organization. A strong foundation will ensure that we are not weak everywhere; but we will not be equally strong everywhere.

> Therefore, determine the enemy's plans and you
>> Will know which strategy will be successful and which
>> Will not;

How to get knowledge about the enemy's plans? Threat intelligence has been talked about a lot in recent years, with many security vendors offering various kinds of solutions in this space. Threat intelligence, as a method for getting advance knowledge of attacks, has been difficult to nail down as a concept or a product. Many vendors are providing hardware or software and selling it as a tool for threat intelligence, when in fact all it is doing is telling us that we have been compromised, in which case it is no longer a threat but an active incident in our infrastructure.

Threat intelligence is many things to many people. My take is laid out above, but threat intelligence is an important tool in many situations. Most of the large IDS/IPS vendors, for instance, have services where, when they see a specific attack play out somewhere in the world, they will update their firewalls elsewhere to detect and mitigate that specific attack. Very clever. That is effective threat intelligence in play. The other kind of threat intelligence, as I see it, is the one companies such as Mandiant conduct, where they dig into the underground and pick up on the chatter going on there, to give customers insight into any new methods or developments in the underground hacker community. Such efforts are very time-consuming and thus a very expensive service to buy. Depending on the size of your company and the threats against it, it may or may not make sense to investigate such a service.

> Agitate him and ascertain the pattern of his movement

All of us cybersecurity defenders have experienced the panic that management, employees, and sometimes even we defenders exhibit during a live attack. When it happens, it is usually an "all hands on deck" situation. While all hands are on deck, other areas like the physical security of the company get put on the back burner, and this provides plenty of opportunities for an enterprising hacker to get physical access to company premises. Or the hacker might sneak in, unnoticed, from the Internet. Both scenarios are equally bad.

How to avoid the almost inevitable panic during an attack? Develop a plan for how to handle a live attack; that way, processes and procedures will be in place for who will be responsible and for how to deal with the attack. When you have good incident response plans in place, the possibility of a full-blown panic will be all but eliminated. Having an incident response plan in place is not enough, however; you will have to exercise it regularly. Without this exercising, the response plan is just that, a plan. It is the regular exercises that will make the plan effective in live situations. To make the plan efficient, make sure to exercise it for different situations, such as ransomware attacks, DDOS attacks, and loss-of-data attacks.

> Probe him and learn where his strength is abundant
> And where deficient

Every single company, organization, and private person around the world is experiencing this probing on a continual basis. It is pretty much a standard MO for a hacker to scan the intended victim before launching any attack. Most firewalls and IDS/IPS tools can alert us to such scanning, because it might well be the precursor to a full attack against our infrastructure.

All of the various scanning tools out there can tell the person doing the scanning what kind of operating system is at the other end, as well as what kind of software is running on the operating system. The grand old man of scanning tools is NMAP, developed by Fyodor back in 1998, but there are other tools out there, most of them free and open source. It is important to point out that these tools also have legitimate uses! Many systems administrators are using them on their own infrastructures to keep track of the number of servers and the software on them.

Scanning tools can provide an awful lot of information to an attacker—how do we limit this amount of information? We can do this by configuring our firewall to ignore certain kinds of traffic, as well as by hardening the servers and software we expose to the Internet. How this is done concretely depends on the firewall and server platforms in use, but spending time doing this configuration is time well spent.

Final Remarks

The takeaway from this chapter, I hope, is the knowledge that without insight into the strengths and weaknesses of both ourselves and our infrastructure, we will be ineffective in our defenses. On top of this, we must have procedures for how to handle any incidents we experience. Without these procedures, the panic which will inevitably happen will be exacerbated to a level where the response efforts will at best become ineffective and at worst be of assistance to the attacker.

To be effective in our defenses, we must know what kinds of systems we have in our infrastructure and which ones an attacker is most likely to target. I have mentioned systems used for storing credit card data as an obvious one, but systems storing personal identifiable information are of equal interest to a hacker. That kind of information can be used to steal identities for various nefarious purposes. Strength is allocating resources to protect the business-critical systems, but strength is also being weak where it is okay to be weak!

The Nine Variables

This chapter and the quotes in it are about how we go about doing our planning and preparation in cybersecurity defense. Without planning and preparation, we will have no effective response to any cybersecurity incident and, without effective responses, such incidents will become increasingly difficult to deal with. Even worse, these incidents will increase in effectiveness and in the damage they do to our infrastructure.

Planning

We make plans for everything in our lives and in our workplaces; without plans, the risks of something going wrong are just too great. So why do we not have plans in place for cybersecurity incidents? A cybersecurity incident will have very real consequences for a company or organization in terms of its ability to conduct business and will therefore have an influence on the ability of the company to make money. On top of that, there are the money and resources needed to recover from the incident.

Having plans, policies, and procedures in place is key to being prepared for the inevitable cybersecurity incident. I have mentioned it elsewhere in this book and I will mention it here as well, because it is key to having effective cyber defense in place. Train in the procedures regularly! Make sure that the policies and procedures are accessible in print form as well—you cannot be sure that the IT systems will be online during a live incident, so having them stored in the infrastructure somewhere will not be beneficial if the systems are offline.

Putting plans in place, with all the moving parts in an infrastructure, is a massive job and not one to be approached in anything less than a serious manner. Because of the agility and rapid change we see within IT today, it is a job which must be repeated on a regular basis and revisited whenever a new change or new piece of software is implemented in our infrastructure. The plans we put in place must be

© Springer Nature Switzerland AG 2019
T. Madsen, *The Art of War for Computer Security*,
https://doi.org/10.1007/978-3-030-28569-2_8

made with a view to the risks and threats we see against our companies and organizations as well.

Quotes

And for this reason, the wise general in his
    Deliberations must consider both favorable
    And unfavorable factors

The wise general in this case is us, the cybersecurity defenders. How many times have we been in situations where some things worked in our favor with other things working against us, in situations where we must secure an infrastructure or a specific system? Dealing with the favorable factors is easy; it is the unfavorable ones that give us headaches. Unfavorable factors come in many forms, and no single solution exist to mitigate them all.

The one positive thing about being in situations with known unfavorable factors is that we are aware of these factors. How many times can you remember being in a situation where you were unaware of critical details, as they related to the security of a system or an infrastructure? In situations where we know of these factors we can put in mitigating controls or, in cases where the situation merits it, decide not to implement a system at all, because of the risks. That last option is only rarely considered seriously by companies. Why? Because in many cases a system gets put into production to support a business process within the company, or the system is designed to make more money for the company by either saving money or acquiring more or new customers.

As cybersecurity professionals, it is our job to identify and point out risks, but it is the business that must make the decision about whether or not the risks identified are within acceptable limits. Acceptable limits differ between different situations and companies. Our job is to make sure the business is aware of the risks and that it is able to make decisions about the risks on a well-informed basis.

By taking into account the favorable factors,
    He makes his plan feasible; by taking into account
    The unfavorable, he may resolve the difficulties

We are, all of us, trying to put plans, technology, and processes in place to secure the infrastructure under our charge. Some of the unfavorable factors might be mitigated by strengthening some of the favorable factors. An example might be a good, strong firewall with an IDS/IPS functionality that is not enabled or used by a company. If an identified weakness can be mitigated by enabling this functionality in the firewall, we will get a stronger firewall in place, as well as mitigate or remove the lack of insight into the traffic running in our infrastructure.

That brings me to one of my favorite points. Before going out looking for a new piece of hardware or software, to help secure an infrastructure, we must investigate what kinds of hardware and software we have running in the infrastructure already. I have seen many times with customers that they already have technology available to them to help secure a system or infrastructure, but they are unaware of it and are only using a subset of the features available to them in the technology. This brings back one of the previous quotes, where insight into ourselves is required for us to be secure. That insight must include insight into the features available to us on the systems running in our infrastructure, as well as which features are unused. That way, we might be able to mitigate risks and weaknesses without having to go out and buy new hardware or software.

> It is a doctrine of war not to assume the enemy will
>> Not come, but rather to rely on one's readiness to meet
>> Him; not to presume that he will not attack, but rather
>> To make one's self invincible

With the level of cybercrime being what it is, we might as well resign ourselves to the fact that an attack on us is a matter of when, not if. Making ourselves invincible is a tall order, and not within the realm of possibility, but we can make sure that when we are attacked, we can identify the attack quickly and recover from it with minimal damage.

Identifying an attack and recovering from it depends not only on us having the right hardware and software in place, but also in equal amount on us having the policies and procedures in place for the recovery. Also, as mentioned elsewhere, we must exercise these procedures on a regular basis. The policies and procedures have no value to the company or organization if they are just stored on the intranet. Procedures only have value if they are exercised.

Final Remarks

This has not been a long chapter, but the information I have tried to convey to you in this chapter is important. Having plans and procedures in place for the inevitable cybersecurity incident is very, very important! Just as important as having these plans and procedures in place is regular training in these procedures and plans. You might get a good review from an external auditor for having these plans and procedures, but the real value of them only manifests itself when the responsible staff are trained regularly and with different scenarios in the procedures. If not, then you just have a bunch of dead trees on a shelf somewhere.

The Nine Varieties of Ground

This chapter will be by far the shortest one in the book and will have only one quote in the quotes section. For this chapter, we see the word "ground" as equivalent to the word "infrastructure." We would like to keep the bad guys out of our infrastructure, and the bad guys want to get into our infrastructure. A simple, black and white situation.

Some areas of our infrastructure are more important to the business than others, and these need better protection than other areas. Do we know which areas these might be? Do we know why these areas are the more important ones? To identify these areas of the infrastructure, we need to know which business purpose or process these areas are supporting, and to do that we need to know about the business and the processes in use in the business. This might sound obvious, but the complexity of current business environments and the various compliance and regulatory require-ments that businesses and organizations are living with makes having a full insight into all these requirements and processes a difficult task and one that, for the most part, cannot be done by just one person alone.

Without this insight into the business, our efforts to protect the infrastructure will at best be ineffective and at worst will contribute to a successful breach of our infrastructure. The days when cybersecurity was an IT issue alone are long past us now. I have mentioned this elsewhere in this book, but I cannot emphasize this enough: cybersecurity is an issue which must be embedded in the highest possible echelons of any company or organization. Without this embedding, resources will not be allocated, nor will the various business unit managers be held accountable for the cybersecurity of business areas. It is this accountability that makes all the difference in cybersecurity; without it, there is a real risk that a laissez-faire approach to cybersecurity will develop among the very people where the responsibility for cybersecurity rests.

© Springer Nature Switzerland AG 2019
T. Madsen, *The Art of War for Computer Security*,
https://doi.org/10.1007/978-3-030-28569-2_9

Quotes

> Ground equally advantageous for the enemy or me to
>   Occupy is key ground

In this case we can switch the word "ground" with "infrastructure." The infrastructure a hacker would really like to get access to is the same infrastructure we would really like to keep the hacker out of. What this infrastructure consists of is not that important, since in most cases it is the data stored in that infrastructure that makes it critical to us, the cybersecurity defenders, and of the utmost interest to the attacker.

How do we go about finding out which parts of our infrastructure contain this kind of data? Data classification. I have mentioned elsewhere in this book that, without us doing regular risk assessments, we will be unable to focus our defenses where they will be most effective for the business. Part of that risk assessment will depend on how we have classified the data stored on the systems under assessment. Now, data classification is a huge subject, and one I will not be discussing much in this book beyond how it ties into risk assessments and sometimes threat assessments. Aside from helping us with the regular threat and risk assessments, data classification will help us configure an effective data loss prevention (DLP) system. Implementing a data classification scheme is not an easy project, or one that should be approached with anything less than a full commitment from the business. IT staff cannot implement a data classification scheme by themselves, because it is the business that knows the value of the data it is utilizing on a daily basis and knows the consequences of this data being missing or incorrect.

Final Remarks

This was the shortest chapter in the book. The takeaway I would like you to have from this chapter is the fact that we absolutely must know the business and the business processes we are trying to protect, if our cybersecurity efforts are to be effective. My remarks on data classification and data loss prevention are certainly important, but including the business unit managers and the C level executives in the chain of responsibility will be the most effective way for any company or organization to take its responsibility for cybersecurity seriously and differentiate the company from its competitors.

With the massive focus there currently is on the security of personal data and privacy because of the EU GDPR legislation, any company that can demonstrate that it is taking these issues seriously and takes appropriate measures to protect its customers will be viewed as a good investment by investors, and customers will be more likely to trust the company with their personal data. In short, cybersecurity is good business.

Attack by Fire

The title of this chapter seems a little aggressive, I know. But, for an attacker, keeping the organization or business under attack stressed and confused is a win, since the staff handling the attack will be unable to see through the "fog of war" to see the real attack unfolding behind all the smoke coming from attacks trying to mask the real attack.

Not every attack is an attempt at masking the real attack. Most attacks will be the real deal, since most hackers will be focused on making the greatest amount of money with the least amount of effort. The kinds of attack which will try and mask themselves behind other attacks will typically be conducted by some form of state-sponsored hackers, since those attackers are more concerned with achieving their goals, whatever those goals might be, than with making money from their attacks.

Quotes

There are suitable times and appropriate days on which
>   To raise fires

In this quote, my intent is to change the wording of the quote to say "attack" instead of "raise fires." We might not always be aware of it, but there are some situations where our infrastructure is more vulnerable than under normal circumstances. During a refresh of the hardware, for instance, or during a company merger. Any time where the infrastructure is in flux, there are opportunities for a hacker to exploit the inevitable weaknesses during that period.

Any time when an organization is experiencing rapid growth or, even worse, downsizing, the processes and structures will struggle with handling the recruitment or firing of employees. Those struggles provide plenty of opportunities for a hacker to exploit the discontent among employees being fired or the stressed HR people trying to staff up the company, by sending them phishing emails or calling

© Springer Nature Switzerland AG 2019
T. Madsen, *The Art of War for Computer Security*,
https://doi.org/10.1007/978-3-030-28569-2_10

discontented employees and trying to get them to act against the best interests of the company.

There is no one single solution here, and a solution that works in one sector will probably be useless in another sector. My recommendation is to be aware of these risks in situations in your company where things are moving rapidly and the IT systems and processes might not be able to keep up with the changes. Any mitigating steps to be taken in situations like these will undoubtedly change from situation to situation.

> Now in fire attacks one must respond to the
> > Changing situation

In this quote, I would like to change the word "fire" to "hacker." All of us who have had to respond to a live hacker attack or virus attack know that the situation will evolve as the hack progresses or we get more information about what exactly is going on. The only way to respond effectively during a live attack is to have procedures in place for how to handle it and to exercise these procedures regularly and with different attack scenarios, as mentioned elsewhere in this book. Aside from procedures, we must have technology in place to tell us about the situation as it changes. Information is the key to keeping track of a live hack as it is happening. SIEM and IDS/IPS systems can help us here, but only if these systems have been configured to keep an eye on the right systems and logs.

What are the right systems and logs, then? That depends on the company or organization. A manufacturing company might want logs from its SCADA systems, whereas a software development company might like to keep track of what is happening in its source control system. Risks and threats are the key components that go into configuring monitoring systems for the needs of a specific company or organization.

Final Remarks

What is the takeaway from this chapter? I hope that you will have gained some insights telling you that not all attacks are what they look like. For us to determine if it is one or another type of attack, we need to have tools which can give us details about the attack and what else is going on in our infrastructure. Log management tools and IDS/IPS systems are the go-to tools here, but just having the knowledge that an attack can be launched against us to mask another and more serious attack will assist us in preparing our responses.

Employment of Secret Agents

Do not get too excited by the title of this chapter. I will not be going for the full "cloak and dagger" with the content here. There will be a little of that, but the focus here will be on how we, as essentially civilian defenders of civilian infrastructures, can use some of the knowledge and experience that various intelligence agencies have acquired since time immemorial.

Knowledge of an attack before the attack is launched is the holy grail of information security. There are many companies and solutions out there promising to deliver in this respect, but you should show care in which solution or vendor you choose. Make sure that the solution caters to your specific needs, in whatever business or industry your company or organization is working in.

Quotes

> Now the reason the enlightened prince and the
> Wise general conquers the enemy whenever they move
> And their achievements surpass those of ordinary men
> Is foreknowledge

We would all like to be notified well before an attack takes place, and threat intelligence promises part of this promised land, but getting this kind of knowledge as a civilian company or organization is difficult, to say the least. We might be able to predict the kinds of attack we might see against our own infrastructure, based on the kinds of attack seen elsewhere in the world, but with the speed of the Internet these attacks might happen on the other side of the world within seconds of being launched for the first time.

So how do we go about getting workable foreknowledge of attacks or techniques that might be used against us? One way is by becoming aware of these attacks while they are still being developed, which neatly brings me on to the next quote.

© Springer Nature Switzerland AG 2019
T. Madsen, *The Art of War for Computer Security*,
https://doi.org/10.1007/978-3-030-28569-2_11

What is called foreknowledge cannot be elicited from
   Spirits, nor from the gods, nor from analogy with past events,
Nor from calculations. It must be obtained from men
   Who know the enemy situation

The enemy here are the very well-organized hacker groups operating from various underground forums and chat channels. Getting access to these forums and channels is in no way easy, as it requires you to build trust with the "community" before getting access, and, even then, you will probably not get access to the entire forum. Most of the big security vendors will have people who are keeping track of these forums, even if they do not have access to the content in there. Those who do have access will be able to keep track of what is being talked about and what kinds of new attack are being developed.

The people who have this kind of access to the underground channels and forums are the people with knowledge of the enemy situation and who can provide us, the defenders, with foreknowledge. That is, if they are willing to share their information. Why wouldn't they want to share? If a company has a person in deep with one of the big hacker organizations, then sharing everything that goes on in that hacker organization would undoubtedly give rise to significant suspicion among the members of the said organization. Just like intelligence agencies, companies with sources in the hacker underground need to protect their sources, especially since the amount of money involved in cybercrime means that any source who is disclosed runs a very real risk of getting killed for his/her efforts.

Delicate indeed! Truly delicate! There is no place where
   Espionage is not used

Industrial espionage has existed for hundreds of years, and for this quote I will focus on the civilian utilization of espionage. For the past decade or so, we have seen in the media various accusations going back and forth between nation states about industrial espionage between nations. This goes on, make no mistake about that; what we as defenders must realize is that our infrastructures are not just the target of opportunistic hackers, but in equal measure a target for the more ethically challenged of our competitors.

In days of yore, an industrial spy needed physical access to a company to rummage through the filing cabinets to find the plans for a company's new product or bids for a job, or whatever a competing company might be interested in knowing. Now, the industrial spy is no longer someone skilled in the art of physical intrusion; now, it is the hacker who has taken over the role of industrial spy, because everything is now digitalized and accessible by an Internet connection.

An industrial spy might also be an insider who has been paid by a competitor to produce plans for a new product. So, industrial espionage might now be done via the Internet connection, but it can just as well be done by a disgruntled employee with an axe to grind or the previously mentioned greedy employee who was paid to do it. How do we defend against these threats? Can we even defend at all against all these threats? No. We cannot provide 100% effective defenses against all these potential threats; what we can do is put in mitigating features, as well as network monitoring to keep track of any sudden huge file transfers from inside the company to IP addresses outside it. On top of that, we must make sure that employees do not have access to files they have no need to access.

Final Remarks

The employment of secret agents brings with it associations of James Bond 007, or maybe Jason Bourne. However cool secret agents may be, though, we should be aware that the professional hacker groups out there, although they are not "secret" in the traditional sense, are not above using a physical person to get access to our infrastructure. By that I mean, for instance, paying one of our employees to get them information, or trying to get one of their own people inside our organization, by tailgating, for instance. Tailgating is walking in just behind someone with legitimate access to the building when they are returning from an errand they have been out for.

The physical security of the various premises that our company or organization is in is just as important to our cybersecurity as the various tools and controls we have implemented. If someone gets access to a laptop and can insert a USB stick, we are toast! With all the outsourcing and cloud computing we are seeing nowadays, our physical security is no longer just a concern for our own premises, but just as much the physical security of our third party providers. Have your policies and procedures accounted for that?

This is the final chapter from the original work of Sun Tzu that I have included in this book. In the next and final chapter, I will try to wrap up some of the information I have tried to convey to you in this book.

The Final Word

Congratulations, you have made it through the book, to this last chapter. I sincerely hope that you have enjoyed the reading and gotten some good tips along the way. In this last chapter, I will try to tell you a little about why I have been so adamant about things like governance, risk assessments, and threat assessments. Governance is the biggest factor in this and covers parts of both risk assessment and threat assessment, so let's start with that one.

Governance

Governance is many things in different contexts. In this context, I am referring to governance as it relates to IT. IT governance is aimed at aligning the IT strategy with the business strategy, to ensure an efficient working relationship between the business and IT. Where does cybersecurity fit into this equation? Cybersecurity is a concern for the business, and not just something that IT deals with. In the current business environment and with the number of breaches we see in the media almost every single week, any serious business will have cybersecurity integrated into its business strategy and anchored in the highest levels of the company.

How you approach governance will be dependent on what kind of business you are in and the threats and risks your business or organization is facing on a daily basis. The kind of business you are in will decide what kinds of compliance and legal issues you will have to deal with. These kinds of issue might also have an impact on the kinds of governance you can be required to implement in your business. I am fully aware that many a geek's eyes will glaze over at the mention of governance—I myself had this reaction for many years—but governance will become the differentiating factor between well-performing companies and companies viewed by both the public and investors as worthy of trust on the one hand, and those which should be viewed with more suspicion on the other hand.

© Springer Nature Switzerland AG 2019
T. Madsen, *The Art of War for Computer Security*,
https://doi.org/10.1007/978-3-030-28569-2_12

Governance will assist with things other than cybersecurity, but in this book I have focused on cybersecurity areas only. Going about implementing IT governance in any company or organization is a massive task, and the IT governance framework must be integrated into any governance already implemented at the business level. If the business has already implemented a governance structure which includes governance for IT, then by all means go ahead and use that structure for your IT governance.

What if the business or organization has no governance structure in place and you would like to implement one for IT? First off, you cannot implement an IT governance framework for IT only. Governance is a business decision, so even if a governance implementation begins with IT, and it often does, the decision and drive must come from the business itself and be supported by the highest levels of the business. Now the question becomes, what kind of governance framework should you implement? There are quite a few of them out there. ISACA has the COBIT framework, which encompasses a lot of other frameworks underneath COBIT. ISO 27001 and CMMi can be included in a COBIT governance implementation, but COBIT is a massive framework and might be overkill in a given business context. I am not going to recommend a specific framework here, but I will recommend that you research a few of them before settling on a specific framework for implementation in your business or organization.

Risks

Risks, along with governance as mentioned above, have been one of the main points in my remarks on the quotes from *The Art of War*. Many governance frameworks cover risks as part of their toolset and with good reason, because risks are the foundation for much of the decision making that any company or organization is doing on a daily basis. Risks differ from one business to another, and the risks faced by a company or organization one year will be different the next year. The risks we see within the cybersecurity realm are evolving at a breathtaking speed, and the old ones are becoming ever more sophisticated in their approach to compromising our infrastructures.

It is risks, along with threats, that we must use in our planning of our cybersecurity defenses. Our defenses must be aligned with the risks to the business we see in the wild; if we do not account for those risks there will be a very real risk, another risk I know, that our defenses will be ineffective in their effort to protect us. Aligning our defenses with the risks has the added benefit that we can use our resources, for which read money, in a way that efficiently mitigates the risks, and not spend money and time on technology or checks where they are not needed. Investing in a brand-new firewall might make sense, but if we have not done a thorough assessment of the efficiency of the "old" one, how do we know we need a new one?

Risk assessment, two words that are easy to put out there, but the effort required to conduct an effective risk assessment is not to be trifled with! On top of that, a risk assessment is not a one-time job; it needs to be repeated regularly to keep up with the changes in the business and in the environment that the business is operating within. This brings me back to the governance points in the previous section. Governance will assist with the checks and balances needed in any serious company, but it will also assist by putting in procedures for regular risk assessments, as well as procedures for risk assessment at the level of individual projects. Risk assessment is something that should be conducted at every level of a business or organization.

Threats

Threats are something a little more specific than risks. Threats are something you can quantify more easily than risks, and you can quantify the consequences of a realized threat with ease as well. Threat assessment has been mentioned in many places under the various quotes in this book and threat assessment, along with risk assessment, is the foundation for any serious cybersecurity defense. Without such assessments and the governance structures surrounding them, our defenses will be ineffective and the resources we allocate to our cyber defenses will be spent in vain.

Like a risk assessment, a threat assessment is not something to go about doing lightly. Like risks, threats are an ever-evolving area of cybersecurity, with old threats being rebottled with new tricks to sneak past our defenses. Threats differ from company to company and from business area to business area, and how we go about defending ourselves against specific threats is dependent on our company and the business we are in with the company. Threats against the IT infrastructure can be identified and dealt with within IT, but threats are not limited to just the IT infrastructure. Threats can be against business processes as well.

The business and, in particular, the business process owners must be included in a threat assessment. Why? Because we as IT people might not know which of our systems are supporting business-critical processes and might thus not consider a specific database or server critical, when in fact the processes these systems support are integral to the profitability of the business. This point is equally relevant for risk assessments. Governance comes into play here as well. Just as for risk assessments, threat assessment procedures and processes are included in any good governance framework. Which is why I began this final chapter with governance.

Final Remarks

This brings us to the end. As mentioned in Chap. 1, I have not been providing you with any specific advice on products or services I think you should implement. I myself am working in cybersecurity using technologies from Microsoft and Cisco, but I have consistently stayed away from recommending any specific vendor here. I have mentioned specific governance frameworks and security standards because these are by their very nature vendor-agnostic. What kind of technology you choose to utilize in your infrastructure should be completely up to you and the business or organization you are servicing.

What I do hope you have taken away from this book is at least one of the quotes and my remarks on it. If you have taken more than one of the quotes to heart, it is even better, and if you have begun to add your own experiences and opinions to the quotes you have read, it is what I would call the pinnacle of what I had hoped to achieve by writing this book. Thank you for reading it to the end!

Appendix: Center for Internet Security Critical Security Controls

There are many kinds of frameworks, aimed at helping companies and users with their security. Some of them are: ISO 2700x, COBIT, COSO and for our purposes here CIS 20. COBIT and ISO 2700x are mostly aimed at the governance parts of cyber security. The advantage of CIS 20 is that this framework is more operational than the other frameworks.

The below list is only listing the 20 major control areas of CIS 20, there are many sub controls beneath each of these major areas, which can assist you in creating and maintaining a secure infrastructure.

Basic CIS Controls

1. *Inventory and control of hardware assets.* Actively manage (inventory, track, and correct) all hardware devices on the network so that only authorized devices are given access, and unauthorized and unmanaged devices are found and prevented from gaining access.
2. *Inventory and control of software assets.* Actively manage (inventory, track, and correct) all software on the network so that only authorized software is installed and can be executed, and that unauthorized and unmanaged software is found and prevented from installation or execution.
3. *Continuous vulnerability management.* Continuously acquire, assess, and take action on new information in order to identify vulnerabilities, remediate them, and minimize the window of opportunity for attackers.
4. *Controlled use of administrative privileges.* This covers the processes and tools used to track/control/prevent/correct the use, assignment, and configuration of administrative privileges on computers, networks, and applications.
5. *Secure configuration for hardware and software on mobile devices, laptops, workstations and servers.* Establish, implement, and actively manage (track, report on, and correct) the security configuration of mobile devices, laptops,

© Springer Nature Switzerland AG 2019
T. Madsen, *The Art of War for Computer Security*,
https://doi.org/10.1007/978-3-030-28569-2

servers, and workstations using a rigorous configuration management and change control process in order to prevent attackers from exploiting vulnerable services and settings.

6. *Maintenance, monitoring, and analysis of audit logs.* Collect, manage, and analyze audit logs of events that could help detect, understand, or recover from an attack.

Foundational CIS Controls

7. *Email and web browser protections.* Minimize the attack surface and the opportunities for attackers to manipulate human behavior through their interaction with web browsers and email systems.

8. *Malware defenses.* Control the installation, spread, and execution of malicious code at multiple points in the enterprise, while optimizing the use of automation to enable rapid updating of defense, data gathering, and corrective action.

9. *Limitations and controls on network ports, protocols, and services.* Manage (track/control/correct) the ongoing operational use of ports, protocols, and services on networked devices in order to minimize windows of vulnerability available to attackers.

10. *Data recovery capabilities.* This covers the processes and tools used to properly back up critical information, with a proven methodology for timely recovery of it.

11. *Secure configuration for network devices, such as firewalls, routers, and switches.* Establish, implement, and actively manage (track, report on, and correct) the security configuration of network infrastructure devices using a rigorous configuration management and change control process in order to prevent attackers from exploiting vulnerable services and settings.

12. *Boundary defense.* Detect/prevent/correct the flow of information transfer between networks of different trust levels, with a focus on preventing the damaging of data (think DMZ here).

13. *Data protection.* This covers the processes and tools used to prevent data exfiltration, mitigate the effects of exfiltrated data, and ensure the privacy and integrity of sensitive information.

14. *Controlled access based on the need to know.* This covers the processes and tools used to track/control/prevent/correct secure access to critical assets (e.g., information, resources, and systems) according to a formal determination of which persons, computers, and applications have a need and right to access these critical assets based on an approved classification.

15. *Wireless access control.* This covers the processes and tools used to track/control/prevent/correct the secure use of wireless local area networks (WLANs), access points, and wireless client systems.

16. *Account monitoring and control.* Actively manage the life cycle of system and application accounts—their creation, use, dormancy, and deletion—in order to minimize opportunities for attackers to leverage them.

Organizational CIS Controls

17. *Implement a security awareness and training program.* For all functional roles in the organization (prioritizing those mission-critical to the business and its security), identify the specific knowledge, skills, and abilities needed to support defense of the enterprise; develop and execute an integrated plan to assess, identify gaps, and remediate through policy, organizational planning, training, and awareness programs.
18. *Application software security.* Manage the security life cycle of all in-house-developed and acquired software in order to prevent, detect, and correct security weaknesses.
19. *Incident response and management.* Protect the organization's information, as well as its reputation, by developing and implementing an incident response infrastructure (e.g., plans, defined roles, training, communications, and management oversight) for quickly discovering an attack and then effectively containing the damage, eradicating the attacker's presence, and restoring the integrity of the network and systems.
20. *Penetration tests and red team exercises.* Test the overall strength of the organization's defense (the technology, processes, and people) by simulating the objectives and actions of an attacker.

You can find more details and recommendations for the controls detailed here on the CIS website at https://www.cisecurity.org/controls.

Printed in the United States
By Bookmasters